Commercial Egg Farming
From Practical Experience Gained Over The Years

by S.G. Hanson

with an introduction by Jackson Chambers

This work contains material that was originally published in 1918.

This publication is within the Public Domain.

This edition is reprinted for educational purposes and in accordance with all applicable Federal Laws.

Introduction Copyright 2017 by Jackson Chambers

Self Reliance Books

Get more historic titles on animal and stock breeding, gardening and old fashioned skills by visiting us at:

http://selfreliancebooks.blogspot.com/

Introduction

I am pleased to present yet another title on Poultry.

The work is in the Public Domain and is re-printed here in accordance with Federal Laws.

As with all reprinted books of this age that are intended to perfectly reproduce the original edition, considerable pains and effort had to be undertaken to correct fading and sometimes outright damage to existing proofs of this title. At times, this task is quite monumental, requiring an almost total "rebuilding" of some pages from digital proofs of multiple copies. Despite this, imperfections still sometimes exist in the final proof and may detract from the visual appearance of the text.

I hope you enjoy reading this book as much as I enjoyed making it available to readers again.

Jackson Chambers

CONTENTS

	PAGE
INTRODUCTION	1
COMMERCIAL EGG FARMING	3
BREEDING	11
INCUBATING	15
BROODER HOUSE MANAGEMENT	19
COLONY HOUSES	27
LAYING HOUSES AND MANAGEMENT	33
CO-OPERATION IN SELLING EGGS AND BUYING FEED	45
ROUTINE WORK	49
MAKING A START	53

LIST OF ILLUSTRATIONS

FACING PAGE

VIEW OF SOUTH SIDE OF 180-FOOT HOUSE 9

INTERIOR OF BROODER HOUSE . . . 19

COLONY HOUSES 27

180-FOOT HOUSE (REAR VIEW) . . . 33

REAR VIEW OF 180-FOOT HOUSE SHOWING YARD 36

INTERIOR VIEW OF 180-FOOT HOUSE . 36

END VIEW OF HOUSES 38

VIEW OF 90-FOOT HOUSE 42

COMMERCIAL
EGG FARMING

INTRODUCTION

Commercial Egg Farming is essentially a a story of first-hand experience and success. It is not dogmatic and is notably free from hobbies. It simply recounts facts as they have occurred. There may be other ways that will succeed but these ways did succeed. And they have succeeded for several people.

The principles set forth in this little volume are as true in America as in England. Its wide circulation cannot but react to the advantage of the poultry industry as a whole, though it should be kept constantly in mind that the distances are greater in America than in England, and that being near a good market is a point of first importance.

In preparing this edition for American readers I have done little beyond expressing costs in terms of American instead of English money. The story remains substantially as Mr. Hanson wrote it.

WILLIAM A. LIPPINCOTT.

Kansas State Agricultural College,
Nov. 1, 1917.

COMMERCIAL EGG FARMING

It has generally been assumed that eggs cannot be marketed at a profit unless their production is carried on as a subsidiary branch of general farming. This assumption rests on the fact that a large number of persons, unfitted by experience and business ability, or with insufficient capital at their disposal have, after a few years' trial of poultry farming, found themselves obliged to discontinue their operations. If an inquiry could be made into the previous business history of any one hundred persons who have given up poultry farming, I believe it would be found that ninety-nine of them had failed in the undertakings in which they had previously embarked, and that many of them had gone into poultry farming as a "last resort." If a man were to go into any kind of business as unequipped for its successful prosecution as the average man embarks upon poultry farming, his end would be equally dismal.

But one does not hear it said on all sides, for instance, that the pawnbroking, the grocery, and the drapery businesses do not pay, as one does of poultry farming, but simply that such-and-such person did not know his business or had not sufficient capital.

It is my opinion, based upon my own practical experience, that poultry farming, purely for egg production, can be made to pay very good interest on the capital invested, if carried out on business lines. My experience has been obtained and a very good income gained from egg production during the last ten years, both in America and in England.

I am not a poultry farmer because I like hens or because I like the business, but simply because I do not know how to earn an equal income so easily in any other way. With me it has been a strictly commercial undertaking. If any other branch of farming had shown me better opportunities I should probably have gone into it.

It may be of interest to those who think of taking up egg production as a business to learn the reasons why I adopted this branch of poultry farming. Until 1904 I was engaged in general farming in Vancouver Island, B. C. The income derived from that industry was earned with difficulty and the prospects were not encouraging. I experimented with various crops and kept careful accounts. The work done was out of all proportion to the earnings. In 1904 I met a man who said he knew something of poultry farming. We made an agreement that he was to furnish the experience and I was to furnish the land and capital. At the end of a year, when he left, my accounts showed a very good return on the capital, though these early beginnings were on the old-fashioned ideas of ten birds in a flock and hot wet mashes. By 1906 I saw that unless I could reduce the labor and hatch on a larger scale, so that a safe income could be assured from egg production alone, my position would not be a much better one than that from which I was endeavoring to escape.

In 1907 I had an opportunity to sell my general farm and risk the money in starting an egg farm. The customary opinion was freely expressed by my neighbors to the effect that no one could make poultry farming pay. The term of my effort was put down at two or three years.

What had finally induced me to embark on the egg farm was that I cleared $1100.00 above cost of feed on 425 hens in 1906. I felt sure that if I could do this with 425 birds, 1000 birds would show, relatively, nearly as good a return.

That first egg farm was a success, though when I started I could not obtain cash for my eggs, but had to get my returns by "taking it out in trade." This of course gave the storekeeper the opportunity of fixing the price on the goods I sold him as well as those he sold me, a condition similar to the one under which most general farmers in America labor to-day.

When I left the small community where I began egg farming, it was, through coöperative effort and following my methods, producing and marketing annually for cash $100,000 worth of eggs.

Though I did not work as hard as my neighbors, all farmers, and I made more money, the labor conditions in British Columbia became acute about this time and there seemed to be no prospect of obtaining labor at rational wages. We were paying $2.43 a day for a Japanese laborer. I found upon inquiry that market prices for eggs in England were such as to show a reasonable profit on their production and that labor could be had at a reasonable wage. So I sold my farm and came to England.

The same dismal prophecies that I had heard in America met me here, when I mentioned that I was going to start an egg farm. With my previous experiences behind me, however, coupled with my knowledge of the climate, soil, and markets of England, I had no fear concerning the results of my methods when applied to English conditions.

I do not wish to infer that my methods are of all methods successful. But the ones described here have during the past several years given me a very good return on the money invested, and others who have adopted this system have secured returns equal to mine.

The following are the most important points in the system as I use it:

1. I breed from constitutionally vigorous hens only, preferably coming-two-year-old White Leghorns, that have shown a production of at least 144 eggs each during their first laying year.

2. I mate these hens to cockerels bred from trap-nested stock, whose mothers have in their pullet year laid not less than 200 eggs each. No cockerel is used which is not of good type and vigorous.

3. The chicks from these matings are brought up on dry feed entirely. No wet mash of any kind is fed to them at any period of their lives. They are never coddled and any chick which shows the slightest signs of weakness is promptly killed.

4. No attempt is ever made to doctor sick birds, young or old. Any bird not looking well is always killed.

5. The pullets are run in flocks of 400 birds each without cockerels, as infertile eggs only are sold for eating purposes. They satisfy my customers better than fertile eggs.

6. My breeding pens consist of 400 two-year-old hens mated with twenty cockerels, all running together as one flock.

7. The housing unit, suitable for 400 birds, is 180 feet long and nine feet wide, which allows a trifle over four square feet of floor space per bird. It faces the south and is surrounded by an acre of land. The south half acre is used in winter and the north half acre in summer. Each yard as it is vacated is plowed and harrowed and sowed with thousand-headed kale. All houses are regularly cleaned and disinfected.

BREEDING

It is a widely accepted theory that constitution is inherited from the female in any animal. Birds, to produce 144 eggs or more in their pullet year, must have strong constitutions and great stamina. It is impossible by breeding from immature females to be certain that the progeny will have strong constitutions. The pullet mother may develop unsuspected weakness before the end of the laying season. She may fail to go through the molt well or may be overcome by heat. Chicks hatched from her in the spring are likely to have inherited her latent weakness. Where one breeds from two-year-old hens he has at least ascertained one valuable fact. The two-year-old bird has gone through her first laying year creditably and has not shown weakness in the molt. If weakness is shown, she is not used as a breeder. Very many poultry farms owe their failure to the temptation to breed from pullets that have not been properly selected or are immature. Though it is a great temptation to use pullet eggs, disaster nearly always overtakes the man who has been so tempted, in two or three years. His stock becomes weaker each year, and less able to produce strong, healthy chicks which will live and stand the strain of laying enough eggs to return a profit.

Too much stress, in my opinion, cannot be put on the fact that constitutional vigor is the first necessity of a successful egg farm. A farmer who knows his job does not breed from a young heifer, neither does a horseman breed from a filly. The reason is the same in both cases. They know that they could not count on getting strong calves or foals.

Having selected vigorous two-year-old hens for the breeding stock, it is necessary to mate these birds with young cockerels bred from hens with trap-nest records of not less than 200 eggs in their pullet year. By experiments it has been proved that the cockerel is much more than "half the flock," so far as good egg production is concerned. The high-producing hen does not transmit her ability to lay large numbers of eggs to her daughters. She does transmit it to her sons, however, and he in turn transmits it to his daughters. It is therefore absolutely necessary to be certain of the production of the mothers of the cockerels to be used as breeders. It is fair that a fairly high price be paid for birds of this type. One pays for the breeder's knowledge, business character, and the expenses incurred in producing birds of high quality. If it is kept in mind, however, that the cockerel is much more than half the flock a good price is but reasonable. Cockerels are used because of their vigor. A two-year-old cock is not likely to throw such strong chicks and cannot take care of so many hens.

The birds must be mated at least ten days before the eggs are collected for incubating purposes. If twenty or twenty-five hens are in one pen, one Leghorn cockerel is sufficient. If fifty females constitute a pen it is necessary to run three cockerels with them, because where only two cockerels are used, one will drive the other away. But one will not drive off the other two, even if he is boss of the pen. With flocks of 100, 200, or 400 hens the number of cockerels is five, ten or twenty, respectively, or at the rate of one cockerel to twenty hens.

As soon as the breeding season is over the cocks (as they will then be) are sold to persons wanting two-year-olds for breeding purposes or sent to market. They are never used a second season on this farm, however good they may be.

INCUBATING

The time to start hatching White Leghorns, bred purely for egg production, is about April 10th. If birds of this breed are hatched earlier than April they will lay too early, possibly in July or August. Then in October or November they will molt and not commence laying again until January perhaps or even February. The period of high prices for eggs is then almost gone.

When I have been short of eggs for my incubators early in the year, I have kept them as long as twenty-one days in a moderately cool room, at a temperature of not less than 40° F. or more than 60° F., turning them slightly every day, and have had fairly good hatches. But the best hatches are obtained from eggs which are not more than seven days old. The eggs to be hatched should be of not less than 2 oz. in weight, and not more than 2½ oz. Very small or very large eggs are not desirable. They should be of good shape, not long and thin or otherwise unusual.

Eggs coming from a distance by rail as a rule will not hatch as well as those produced at home and placed in the incubator without jarring. But as I find it necessary to buy eggs for the purpose of obtaining cockerels to mate with the following year's breeding hens, the loss entailed by broken eggs and broken yolks must be borne with. I have made it a practice, during all the years in which I have been poultry farming, to introduce new blood from the very best flocks in existence, and the wisdom of this practice is proved year by year by the continuously increased egg average of the flock, the greater stamina, and the better health of the birds.

There are many who believe that the introduction of new blood into a flock from even the best outside sources will only do damage as far as egg production is concerned, and consequently these breeders practice "line breeding." Line breeding is only possible for the expert breeder, for if mistakes are made, all the mischief of inbreeding must

INCUBATING

soon be apparent in the flock. These are usually low fertility of the eggs, difficulty in rearing the chicks, and loss of stamina in the birds. My experience has proved to me that it is not necessary to "line breed" in order to keep up or increase the egg average, though it is well to purchase cockerels from those who do.

All incubator manufacturers furnish rules with their machines, so that nothing need be said on the actual running of an incubator. It is necessary with all machines that the room in which they are placed should be well ventilated, and even in temperature. I now use a "Mammoth Incubator" with a capacity of 2500 eggs, but the management is just the same as for a small one. Formerly I used six incubators, each with a capacity of 400 eggs, and in consequence had to attend to six lamps. Now I have one heater which burns anthracite coal. The machine is not cheap, but it saves an hour a day in labor and the cost of running it for the same number of eggs is less than when I used oil.

BROODER HOUSE MANAGEMENT

The perfect brooding arrangement has not yet been invented. All systems have one or more defects. The cold brooder is not suited to the English climate nor to most parts of America. Oil-lamp brooders are difficult to ventilate properly, and the fumes from the oil are dangerous. Brooder houses heated with hot-water pipes are the most economical to run, but the per cent of deaths among the chicks is considerable. The large American brooder houses, 24 feet by 12 feet, for flocks of 600 chicks, heated by a central stove, when in charge of an experienced person, are said to give good results, but I have had no experience of their management.

I started, as I have said, with oil-lamp brooders, but the lamps and the ventilation of the sleeping quarters required so much attention that I saw that if large numbers of chicks were to be raised some other system would have to be adopted. People who are only raising 200 or 300 chicks can manage with oil lamps. But when one goes into poultry farming commercially, chicks must be raised by the thousand at the least cost of time and labor. The brooder house which I use is 110 feet long by 12 feet wide. There is a four-foot alleyway running along the back. The front is divided into twenty compartments, five feet wide, each with a capacity of 125 chicks. Ten compartments are on either side of the stove, which is in the center of the building and occupies a space ten by twelve feet. The stove is an ordinary water-jacket arrangement with fairly large coal capacity. Two two-inch pipes run from the stove to each end of the building and return. Ordinary anthracite coal is used as fuel. If the fire is properly made, it will last for twelve hours without attention, and will keep the water in the pipes hot enough for quite young chicks. The pipes start from the stove at about five inches above the floor level and gradually rise at the rate of five inches in each fifty feet, which gives a good circulation. The floor of each

BROODER HOUSE MANAGEMENT 21

compartment is sprinkled with coarse sand, and over that is a layer of chaff. As the pipes rise, more sand is put on the floor underneath them to bring the floor level up to within three inches of the pipes. A board is placed over the pipes to reflect the heat down on the chicks, and for the first week a sack is thrown over the board for extra protection.

About a week before the chicks are due from the incubator, the fire in the brooder house is started and kept going in order to thoroughly dry the sand, which is spread in the pens some months before. When the chicks are well fluffed out in the incubator, they are placed in boxes and taken to the brooder house and put under the pipes. They are allowed to run about near the pipes, the side nearest the windows being blocked with a board for two days, by which time the chicks have usually learned the location of the source of heat. The board is then removed, and they have the run of the whole compartment. When first allowed the full run of the pen, some attention must be given to seeing that all the chicks find their way back to the warmth. If it is noticed that any are standing about or crying they must be put under the pipes. This must be continued until it is certain that all know their way back.

The first feed is given when the chicks show unmistakable evidence of hunger, which is thirty-six to forty-eight hours after hatching. It consists of "kibbled" wheat or any ordinary commercial chick feed. If it is the latter it need not contain grit. The coarse sand on the floor is quite sufficient for the chicks in the way of grit. Water, in which there is enough permanganate of potash to make it wine red, is given with the first feed of grain, and thereafter during the chicks' life in the brooder house it is given twice a day. The drinking vessels on each occasion are thoroughly cleaned and rinsed with disinfectant. This is most important. It is nearly impossible to keep water for chicks clean, and the cause of many chick troubles can be traced to an insufficient supply of perfectly clean water.

When the chicks have the full use of their brooder house pen, sufficient grain is scattered in the chaff to last for two days, and a little ground charcoal is given in a tray. This is continued during the first week. The charcoal seems to help keep the bowels in order and prevents many chick disorders.

During the second week the sacks are taken from the boards over the pipes, and a dish containing bran mixed with charcoal is placed in every pen. Enough grain is fed to last two days and water is given as before. If the weather is reasonably fine the chicks are allowed in the outside runs. There they find green feed in the shape of kale, turnips or lettuce. During the first day of two of their new experience, fairly constant attention must be given to see that all are able to find the way home. They get chilled easily and bowel trouble ensues.

The third week dry mash is given in place of the bran, this being the only change in the second week program.

During the fourth week the proportion of grain is increased somewhat. The board is removed from the top of the pipes, and a small door is opened at the end of the outside run which admits them into a half-acre field, where the chicks of one age mix. They may find their way back to any pen that suits them. Some pens will have twice the number of chicks that others do. I think that the extra exercise which this arrangement allows does the chicks more than enough good to overcome the harm that comes from a little crowding in some of the pens.

This in brief is all the attention my chicks receive while they remain in the brooder house, and they are thrifty and strong. I must repeat, however, that any chick which shows signs of weakness or evidence of disease is promptly disposed of. A chick is not of great value, but it may be the cause of great loss if it harbors disease.

A good many cockerels can be distinguished at six weeks of age. Some are sent to London and bring from eighteen to twenty-four cents each. Others are kept for another month or so and sold for about thirty-six cents. There is no money in fattening Leghorn cockerels in England. It is therefore better to sell them for what they will bring rather than give them time, labor and house accommodation.

BROODER HOUSE MANAGEMENT 25

In America there is usually a good market for Leghorn cockerels of about ten weeks and weighing a pound to a pound and a quarter, though the prices vary greatly with the different sections of the country.

Where the prevailing prices are satisfactory the cockerels intended for market are put on a forcing ration as soon as their sex can be distinguished. The forcing ration differs from the ordinary growing ration largely in its increased proportion of ground feed. The cockerels are continued on this ration until they are nearly the size desired when they are put in crates which limit their exercise and are milk fed. The milk fattening ration may consist of:

> 2 parts corn meal
> 1 part wheat middlings or shorts
> 1 part oat flour
> 8 parts buttermilk

I always save cockerels for breeding purposes, the basis of selection being vigor, There is a house with alternate yards for cockerels only, where they are kept over the autumn and winter, running together as one flock. No attempt is made to keep them from fighting, the best birds among the survivors being used as breeders the following season.

The runs outside the brooder house are dug up as soon as they are vacated. They and then limed and seeded to some quick-growing crop. If possible two crops are taken from these yards before the next spring so as to insure the sweetening of the land. Though these yards are used but eight weeks in the year these precautions are religiously carried out every year. If evidences accumulated of the land becoming "chicken sick," or full of gape worms I should disinfect the yards with a strong disinfectant before digging them. If that failed I would remove the soil to a depth of six inches and replace it with absolutely clean soil.

As soon as the brooder house is emptied the pens are cleaned and very carefully and thoroughly disinfected. All the boards are creosoted and fresh sand is put in the pens to be in readiness for the following spring.

The cost of coal for this brooder house works out at about a half cent per chick per month.

COLONY HOUSES

When the chicks are about six weeks of age they are taken out of the brooder house if the weather is reasonably good. If the weather is cold and wet, they are kept in the brooder house another week or more. The arrangements for ventilating and heating in the brooder house are such that the chicks always get plenty of fresh air, and can always get away from the pipes if they feel too warm. The danger of too much heat, which is weakening, is thus reduced to a minimum.

The colony houses are only used until the pullets are ready to go into the laying houses. Any style of house will serve provided there is plenty of air and the floor is dry. I use a sort of miniature laying house, four by five feet with glass in front. The air comes in under the hood as in my laying houses. Three perches can be put in, and a detachable arrangement, which looks like a nest box with a hinged lid from the outside, but which is divided into three compartments to hold dry mash, grain, and water, is placed under the window in the front of the house. The floor is covered to the depth of two inches with peat moss litter. Wire netting is put up around each house to confine the birds until they are accustomed to their new home. Generally after they have been confined in the yard for two days the netting can be rolled up, and the birds given free range. They usually find their way back all right.

They are watered twice a day. Grain, dry mash, and coarse sand mixed with grit are always kept before them. If the summer is dry, extra green stuff is given. The houses are cleaned once or twice a week as required, and once a month they are disinfected by means of a spray pump and a strong disinfectant.

When the birds are first put in the colony houses, ordinarily fifty in each house, they have a tendency to crowd together for the first few nights. To prevent this I make up the corners of the houses so that they are raised above the ordinary floor level. (I am told that if a stable lantern is hung from the roof, so that the bottom is two or three inches from the floor, the birds will collect round the light and not crowd, though I have not yet tried it.) Besides losing birds by suffocation, the crowding interferes greatly with the general thrift of the flock.

I differ from many others in the matter of allowing birds to perch early. I am always pleased when they take to the perch, because crowding troubles are then at an end, and I have not found that my flock has crooked breast bones in excess of the number in other flocks which are not allowed to perch as early as mine.

The birds stay in the colony houses until they are about five months old, when they are moved to the laying houses. The latter have gradually been emptied of those birds whose life of usefulness, as far as I am concerned, is at an end. I endeavor to restock half my houses every year with pullets, the the other half containing hens from which I breed the next year's pullets. I believe that it would be better to restock three-quarters of the houses every year, as far as the income from commercial egg production alone is concerned, using only the remaining one-fourth for breeding stock. But I have been fortunate enough to secure a market for eggs for hatching, which takes during two months of the year all the eggs which are not required for my own hatching purposes. I therefore find it necessary to keep over a large number of hens.

There are many who argue that the cost of raising so many pullets yearly is too great, and therefore they keep the hens another season and raise only one-third of the total stock each year. In my opinion this is not sound business. If I were short of stock because my hatching and brooding season had not gone well I would without hesitation keep hens on for another season, which in ordinary good breeding seasons I would sell in August and September. But I know from many years' accounts that my net income would not be as good as if I had my usual supply of pullets. For one thing, it is very unusual to get eggs before January, from even one-year-old hens after the molt, and then not in paying quantities. But pullets will be laying some in November, better in December, and very well in January. These are the months of high prices for eggs. One dozen eggs during that period is worth two or more dozens in the spring.

Taking one year with another, the cost of raising pullets to the laying age, including the value of the eggs for hatching and the cost of fuel, labor and feed, is about offset by the price which I get for the hens wholesale in the London market at the end of their second season. From my standpoint then it would be foolish not to raise yearly all the pullets which I require to restock half the houses. I believe that experiments have proved that one dozen pullets as profit-earners are equal to two dozen yearling hens, and to three dozen two-year-old hens. That, I think, is pretty nearly my own experience.

LAYING HOUSES AND MANAGEMENT

The houses which I use are 180 feet long by nine feet wide; seven feet high in front, four feet high at back. They will comfortably house 400 birds, which allows a little over four square feet per bird. Each house is situated in the center of an acre of land. One half-acre is on the south side and is used in the winter. The other half-acre is at the back of the house and is used during the summer. Small doors, both in the south and north sides, give access to the respective yards. By these means one-half the land gets a rest from the birds for six months out of every twelve, and during the period of rest is plowed, harrowed, and a green crop sown. By the end of the six months there is plenty of green stuff for the birds to eat, and the growing of the crop has sweetened the soil. It is necessary to either move the houses every year or to arrange as I have done, and give the birds practically fresh land. Where the houses are moved only small houses can be used, and it is the small-house system which I want most to avoid because of the extra labor involved where the number of separate flocks is increased. It does not matter how small a flock is, or how big the area of land on which the birds run, in time the land around the house will become fouled. It is therefore simpler to arrange alternate yards, which are cultivated and cropped in turn. Merely digging or plowing is not sufficient in itself. Crops must be grown to remove the manure, and the birds should be kept off that part which is under crop, for the full period of six months.

The advantages of running fairly large flocks of birds in a big house are these: fewer doors to open and shut, less wire-netting, fewer drinking vessels to clean and fill; and the yards may be plowed rather than dug by hand. At first sight these advantages do not appear great, but suppose that instead of my four houses, holding 1600 birds, I had

LAYING HOUSES AND MANAGEMENT 35

160 houses, each holding ten birds. That would mean 160 doors to open and shut at least three times a day, as against opening and shutting eight doors for my four houses. The difference in time in one day would be considerable. Roughly speaking, what is done in a week's work on my farm in regard to this item alone has to be done by the small-house system every day. The same applies to the number of gates going into the yards.

Then there is a great saving of money in the netting for the yards. These 160 houses would require two runs each, and these would require about 5280 yards of wire-netting with posts every ten feet. My eight runs, which accommodate the same number of birds as the above 320 runs, require only 2200 yards of wire-netting with the proportionate number of posts. Plowing, which can of course only be done in big yards, is much cheaper than digging, which must be resorted to in small yards. Each of the small houses must have a water dish, in all 160 dishes. I have eight dishes for each house, thirty-two in all. I estimate that I save on these items not far short of four hours a day of one man's time.

The other advantages of the big-house system are that the attendant is under cover in bad weather, when feeding the birds, collecting eggs and cleaning the houses. The chief advantage as far as the birds are concerned is that they have a large, dry, well-ventilated place in which they are fed and can scratch, and are thus independent of the weather, however bad it is.

It is sometimes claimed that small flocks give a better egg yield per bird than large flocks. This is not my experience. For three years I used various types of small houses, holding from ten to fifty birds, and never had as good results as I now have in my big houses. Possibly I may know more about management and breeding than I did in the early days, but nothing that I have seen or read has shown me that the extra labor and expense involved in the small-house system is compensated for by an increased egg yield. It is also claimed that there is less danger in the small-flock system should disease break

LAYING HOUSES AND MANAGEMENT 37

out. I can believe that this may be so, but I maintain that if the methods I follow are carefully and properly carried out disease will not appear in the flock. To make profits is the sole object of my poultry farming, and the big-house system has been the natural outcome of my experience.

My houses are divided into compartments ten feet wide. The partitions run six feet across the house. The object of these partitions is to keep draughts from striking the birds when they are on the perches. The perches are at the back of the house, raised nine inches above the dropping boards, the latter being two feet above the floor. The front of the house is boarded two feet six inches up from the floor. Above the boards is glass. In Canada, although much colder than in England, I used wire-netting only, but in my present locality the winds are so strong that when there is rain it is blown clear across the floor. Therefore I put in ordinary horticultural glass, held in place by strips of wood, to the level of the bottom of the hood. The hood projects about twenty inches and admits all the air that is necessary. A proof of this is that at no time of the year is there any objectionable odor in the house. The roof, back, and sides of the house are covered with a good roofing felt. If a house has a large open ventilating space, as my houses have, it is necessary to make the back of the house as nearly airtight as possible. Boards, however good and closely fitted, are bound to allow some draft. Fresh air without draughts is necessary if colds are to be avoided. Nest boxes are placed in front of the house, and below the level of the glass. They are so arranged that they can be taken off and easily disinfected. Above and close to the nest boxes, and directly under the hood, are a few broody coops. The hood and the front of the house form two sides and the bottom consists of strong wire-netting. There are always a few broodies, even among Leghorns, but these airy broody coops soon cure them of the inclination to set. The floor of the house, which is of wood, is covered with two or three inches of peat moss litter or dry sand, above which there is six inches of straw. Fresh straw is put in whenever the old becomes broken in small bits.

LAYING HOUSES AND MANAGEMENT 39

Only once a year is the litter completely cleared out, at which time the floor is tarred or creosoted before the fresh peat moss or sand is put in. The houses are so perfectly dry that it is not necessary to clean out completely more often. The dropping boards are cleaned twice a week, and in spring, summer, and autumn are disinfected once a month by means of a spray pump throwing a strong commercial disinfectant. At the same time the back of the house, both above and below the dropping boards, and the partitions, are treated in the same way. The perches are taken out of their sockets and creosoted every month, except in winter, when paraffin is used because it dries quicker.

My houses, being 180 feet long by nine feet wide, have a floor space of 1620 square feet, which gives four square feet per bird when 400 birds are in one house. The grain is fed in the morning in the litter on the floor, at the rate of about twelve pounds per 100 birds. Whatever the state of the weather the scratching in these six inches of dry litter promotes exercise and keeps the birds healthy. Little should be expected from a hen when she is fed her grain on sodden earth outside a dark roosting house, in the rain and wind. In the worst weather the hens on this farm can be seen scratching all day long in these dry, well-lighted, and well-ventilated houses, whereas on similar days the birds housed in small houses would be humped up miserably in a corner or sitting on a perch. These houses before the war cost about five hundred dollars each, that is, a dollar and a quarter per bird. They last for very many years, the only upkeep being the annual creosoting of the inside and front outside, and every other year a coat of tar for the roof and the back outside.

Every partition has a dry mash box fastened on it, the bottom of the box being about nine inches above the level of the litter on the floor. The boxes are twenty-four inches long, eight inches wide, and six inches deep. Laths are nailed at intervals of three inches across the top to prevent the birds from getting into the boxes and scratching the mash out. The boxes are never filled more than three-quarters full, as the birds would hook the mash out with their beaks.

The advantages of feeding dry mash are found in the greater health of the birds, and the saving of labor.

To mix wet mash and feed 400 birds takes at least one hour a day. Unless the mash is carefully mixed to the right consistency it is likely to cause bowel trouble. In addition, it is not always possible to judge the amount which the birds will eat on any particular day. Consequently on some days they may get too little, and it is necessary to mix more, or too much may be given, and then it is wasted. Further it must be gathered up, because once wet it will not keep. Dry mash is always accessible to the birds, the boxes being filled twice a week. If I had more or larger boxes I would only fill them once a week. As it is dry and in a dry place, it never spoils. One may be sure that every bird gets as much as she wants. There is no crowding, as in wet-mash feeding, with the result that the weaker birds get their proper share. There is no gorging, because it is impossible for a hen to take more than a few mouthfuls at a time on account of its dryness. She then has to go for a drink.

Then she comes back for a little more mash, and then goes for a little more water. Thus, with having to scratch for grain, moisten her own mash, and go out in the yard for green stuff is she kept busy. A busy hen is always healthy, and more likely to be in condition to lay. The wet-mash fed bird will often swallow a lot of mash in a few minutes and then loaf about in a corner. The result is she puts on fat and does not lay. Again, the dry-mash feeding is much the best for any-one starting poultry farming, as the knowledge required in properly wetting mashes and feeding them is considerable. The dry mash is left for the birds to help themselves, and the moistening rests with them.

The composition of dry mash varies with the price of the feeds, and slightly with the season. The proportion of nutrients that I try to obtain for the dry mash is always such that when combined with the grain fed in the morning the nutritive ratio is one part protein to a little over four parts carbohydrates.

I usually use ground oats and soy bean meal in my mash, but at the present time both these feeds are too dear for my pur-

LAYING HOUSES AND MANAGEMENT 43

poses. In consequence I am using a mash composed of 50 lbs. of bran, 50 lbs. of middlings, 50 lbs. of maize meal, 50 lbs. of fishmeal, and 1 lb. of salt, with cracked corn fed as grain in the morning. I am getting very good results in the matter of eggs.

Cracked corn is the cheapest grain at present (January, 1916). It has a name for being too fattening and causing liver trouble. I have, for my part, never found any harm in feeding corn in any season. In early days I fed whole corn only, no other kind of feed whatever being given to the birds, in all temperatures, from 3° below zero to 100° in the shade, and I had a flock that averaged 178 eggs per bird. I do not know whether cracked corn would do with the heavy breeds, but I know it gives good results when fed to White Leghorns.

As I said before, the birds are bred from when two years old, and are mated with cockerels bred from hens with high egg yields. About twenty cockerels are placed with 400 hens, and the fertility has always been high.

The best of the hens are sold at the end of

the breeding season for breeding purposes, the others are sent alive to market.

That is the whole system. It is simple, and I hope I have made it clear.

COOPERATION IN SELLING EGGS AND BUYING FEED

There are many firms in London and the larger cities of America which will give a better price for a regular supply of eggs of standard quality and size from a single person than for irregular supplies from several farms. Both in America and England I have proved that a better price is obtainable when large quantities of eggs are for sale. The chief points in selling eggs are to be sure that they are new-laid, infertile, clean, and not less than 1⅞ oz. in weight and two ounces is better. The smaller eggs usually can be sold to the same firm at a slightly lower price. The railway companies of England will make a lower rate than usual for eggs sent in large quantities, such as two cwt. or three cwt. at a time.

It has not been found necessary to depend on the coöperative marketing of eggs to so large an extent in the States as in England. Where it is necessary, however, the following matters should be taken carefully into account. All members of a coöperative egg marketing society must be persons who can be relied upon to keep the rules of the society strictly. This means that their farms must be run in accordance with these rules, and that the farms are open to inspection by the secretary without notice. One of the rules must be that eggs may not be bought by any member from a non-member and sold to the society as produce from his farm. The reputation for supplying really fine eggs can be lost by carelessness in any one farm. Therefore one of the most important points is the character of the persons who are members. A member must be a man who can be relied upon to observe all the rules faithfully. It would be impossible to admit as a member a farmer running his birds as farmers generally do, that is, on what is called "free range," which means eggs laid in hedges, under stacks, in stables, etc., and found many days or weeks, perhaps, after having been laid. The temptation to sell them as "new laid" is very often too great to resist when prices are good. Hence "farmers'" eggs do not fetch the top price in any city in the world. Again, the farmers' hens have access to manure heaps and eat all kinds of disagreeable things. The flavor of an egg frequently depends on the way a bird is fed. To prove that this is so, try feeding hens on onions, fish, or even rape, and test the flavor of the eggs.

The advantage of buying feed in large quantities must be apparent to everyone. Capital is not needed where a small society is formed, every member of which pays cash within fourteen days for the feed ordered through the secretary. In our small society here we have been able to buy as cheaply as the local corn merchant.

ROUTINE WORK

The routine varies with the seasons; but, roughly, any week's work on my own farm is as follows:

FROM OCTOBER TO MARCH

Monday morning: Feed grain, clean and fill water dishes, mix about 800 lbs. of dry mash and replenish dry mash boxes; clean, sort, and pack eggs; dispatch eggs to London. Afternoon, clean and fill water dishes; collect eggs; shut houses.

Tuesday morning: Feed grain and give water as on Monday; clean all dropping boards; once a month, after cleaning, disinfect all houses; clean, sort, and pack eggs. Afternoon, water birds; collect eggs; shut houses.

Wednesday morning: Feed grain and give water; clean, sort, and pack eggs. Afternoon, water birds, collect eggs, and shut houses. If fresh straw is wanted in the houses it is usually put in on this day.

Thursday: Feed grain; water birds as on other days; clean, sort, and pack eggs; dispatch eggs to London; mix dry mash, 800 lbs., and feed as on Monday; collect eggs; shut houses.

Friday: Same as on Tuesday, except that once a month creosoting perches is done on this day.

Saturday: Same as on Wednesday; in addition replenish grit and shell boxes.

Sunday: Same as on Saturday as regards feeding, watering, collecting eggs, and shutting houses.

APRIL TO MAY

In addition to attending to the hens as mentioned in the previous routine week:

Morning: Attend to incubator fire; turn and cool eggs; attend to brooder house fire; water chicks; feed chicks.

Afternoon: Attend to incubator same as

in morning; shut brooder house doors. Ten P. M. make up incubator and brooder fires for the night.

JUNE TO SEPTEMBER

In addition to attending to the hens as usual; attend to the brooder house (during June) as mentioned above; open colony house doors; attend to pullets; water; feed grain, mash, etc.

In August completely clean out all litter from laying houses, tar or creosote the floor, dropping boards, perches, and nests; put in dry sand or peat moss dust, and six inches of straw; then move in pullets, as the hens cease to lay, and are sent to market.

MAKING A START

If the prospective poultry farmer has a house, outbuildings, and eight or ten acres of land, he will need in addition at least five dollars per bird capital, which means that if he intends to run a farm of 1000 layers, he must have $5,000.00 cash and sufficient means to live on for two years. If he intends to run a farm of 500 layers, $2,500.00 cash capital will be needed.

The dwelling house, land, and outbuilding are not considered as part of the capital for poultry farming, but as necessaries before an investment can be made. It would not be fair to charge the poultry farm with the rent of the dwelling house or garden, but the rent of the land and outbuildings must be debited to the poultry farm. As these vary so much in different localities, it would not be possible to state a definite sum.

It is to be clearly understood that the five

dollars per bird capital is quite apart from the other charges which I have mentioned, and the poultry farmer must see that he has a balance in hand in the proportions mentioned above, after paying for his residence, land, and outbuildings.

The cheapest way to start is to build a laying house by October, and buy sufficient good birds to fill it. The number of birds (and consequently the size of the house) must correspond with the number of eggs which will be required for hatching in the following spring. Thus, if 2500 eggs are to be incubated in two hatches of 1250 eggs each, and the oldest of the eggs are not to be more than seven days old at the date of starting the incubator, it will be necessary to have as many hens as will lay 1250 eggs in seven days. Seven into 1250 goes 180 times. Allowing for a margin of eggs which are not suitable for incubating purposes, 200 eggs a day will be needed. Supposing that March 21st is the day to start the incubator, the two-year-old hens should be laying at the rate of 66 per cent per diem at that time of the year. So to get 200 eggs a day 300 two-year-old hens must be bought. These birds will hardly pay for their feed during the winter. The benefit will be derived in the spring, as eggs laid on the farm will hatch better than eggs that are shipped in. The eggs from these birds after the hatching season can be sold commercially and will show a profit over the cost of feeding. The birds can be used another season in the same way and then sold in the autumn. The first year's pullets will then in the following (third) spring be old enough to breed from.

Other ways to start are to buy eggs for hatching in the spring, or to buy day-old chicks. The latter way is more expensive though it saves incubating labor and, temporarily, the cost of incubators. If the farm is to be run economically the latter expense must be faced sooner or later. In my judgment day-old chicks should be bought only if it is impossible for one reason or another to do the incubating.

A proper incubating room must be ready by February 15th, and incubators installed by March 1st, so as to allow time for running the machines empty, before filling them with eggs. The brooder house must be ready by April 1st. This will allow time for learning how to fire and to dry the house before the chicks are put in.

Assuming that the farm is to have a capacity of 1000 laying hens, it will be necessary to raise 500 pullets to the laying age during the first year. To make certain of this it is necessary to incubate 2500 eggs. Two thousand eggs may do, but it is safer to allow a margin, as the period of the year during which eggs can be hatched to produce pullets which will lay at the right time is limited from March 15th to April 21st. If from any reason bad luck should attend your efforts, say, about June, and chicks are lost, then the margin I have mentioned will carry you over. If bad luck does not come it is easy to dispose of surplus well-bred pullets. Usually too few are bred, rarely too many.

The margin may to many seem large, but on big American farms 8000 eggs are frequently incubated to produce 1000 pullets at the laying age. Allowing 10 per cent for infertiles at first test from 2500 eggs leaves 2250. Allowing 10 per cent for dead germs

MAKING A START

at the second test from 2250 eggs leaves 2025. Allowing 60 per cent chicks from remaining 2025 eggs gives 1215 chicks. Figuring 50 per cent cockerels, leaves 607 pullets and allowing for 15 per cent deaths, leaves 517 pullets at laying age.

These 2500 eggs from the right kind of stock will cost about $450.00. Feed for the pullets to laying age will cost about sixty cents per bird, giving a total of three hundred dollars. This is figuring on the basis of using 7000 pounds of grain and 7000 pounds of dry mash to rear the 500 pullets to the age of six months. The cockerels can be left out of account. Reckoning the cost of the hatching eggs, incubating and brooding charges, the selling price of the cockerels is usually about equal to their cost of production.

Incubators for hatching 1250 eggs at one time will cost about $135.00. A room suitable for the incubators may be found in the outbuildings if fifty dollars is spent on it. A brooder house to hold 1200 chicks costs about $500.00 with yards. A laying house to hold 500 pullets will cost at present about

$1000 with yards. (Before the war it cost about seven hundred dollars.) So that by October the following sums will have been spent:

Eggs for hatching	$ 450.00
Feed	300.00
Incubators	135.00
Incubator room	50.00
Brooder house	500.00
Laying house	1,000.00
	$2,435.00

For drinking vessels, etc., allow five dollars.

A granary will be necessary. If an old building can be used fifty dollars might fit it up. A pony, costing fifty dollars; a manure cart, a dry-mash cart, each costing, say, twenty-five dollars, wheelbarrow, scrapes for cleaning boards; fifteen egg boxes, each holding thirty dozen eggs, cost two dollars each, total thirty dollars. An egg-packing room is also necessary.

MAKING A START

SECOND YEAR

Eggs for hatching	$ 450.00
Feed for chicks	300.00
Another laying house	1,000.00
Small cockerel house	125.00
Egg boxes, pony, etc., as mentioned above	200.00
	$2,075.00

A water system will have to be provided; a well with three H.P. engine and pump, tank, and stand pipes, etc.

The income from the first year's pullets, above the cost of their feed, should be $750.00 and might be more.

THIRD YEAR

The first year's pullets are now two-year-old hens. From these birds one should breed 500 pullets, and then sell the breeders as they stop laying in the autumn. Sufficient colony houses should be built so as to be ready in June. These will be occupied by the 500 pullets until the laying houses are emptied of the hens. The cost of these colony houses will be about two hundred dollars for twenty houses if the cheapest grades of lumber are used.

The income from the farm, in the third year, should be $1500.00 or more on the $5000.00 invested.

If hens are bought in the autumn for the purpose of obtaining hatching eggs in the spring, it is well to buy those which are about eighteen months old. For such birds, if of good strain and from a reliable breeder, a dollar and a half to two dollars each must be paid. Cockerels to mate with the hens will cost three to four dollars each.

The money to be spent, if hens instead of hatching eggs are bought to make a start, will be:

FIRST YEAR

300 hens at $1.50	$ 450.00
15 cockerels cost	50.00
Laying house and yards for 315 birds	725.00
Feed for 315 birds for 12 months (14,000 lbs. grain 14,000 lbs. mash)	500.00
Incubators	135.00

Incubator room	$ 50.00
Brooder house and yards	500.00
Laying house and yards for 500 pullets	1,000.00
Feed for growing stock (7000 lb. grain, 7000 lbs. mash)	300.00
	$3,710.00
Less sale of eggs not wanted for hatching, and cockerels	750.00
	$2,960.00

SECOND YEAR

Feed for hens (14,000 lbs. grain, 14,000 lbs. mash)	$ 500.00
15 cockerels cost	50.00
Feed for growing stock (7000 lbs. grain, 7000 lbs. mash)	300.00
Enlarging the house holding 300 hens to hold 500	375.00
Small cockerel house	125.00
Egg boxes, pony, etc., as in former account	200.00
	$1,550.00

CREDIT

Sale of 300 hens at
fifty cents.........$150.00
Sale of eggs from 300
hens 500.00
                            ———
                            650.00
                            ———
                            $900.00

The income from the pullets will be the same as in the former account.

If these figures are taken as approximately accurate, the first method has cost:

1st year: total expense........$2,435.00
2nd year: total expense........ 2,075.00
                                ———
                                $4,510.00

The second method has cost:

1st year: total expense........$2,960.00
2nd year: total expense........ 900.00
                                ———
                                $3,860.00

The second method has cost, therefore, $650,00 less than the first method, but has involved six months more labor.

I repeat that these figures must be taken to be approximate only as the market quotations for labor, building materials, and feeding stuffs vary from year to year. At the same time, the estimates of profit, based on ten years' accounts are approximately correct.

A 90-Foot House to Carry 200 Hens to Produce 16,000 to 20,000 Eggs Per Annum.

Ends of Houses. Facing East and West Always. R.O.K. Felting Over Ends, Including Doors. Gate for Pony and Cart. No Hinges; No Iron Bolts. Pullets in North (Side) Yard; Summer.

180-Foot House (Backs) Showing North Side of Each House, with the Half Acre of Land Planted with Winter Vetch, in Use During the Warm Months—April to Early October.

Colony House.

Colony House. For Chicks from Eight Weeks Old to Six Months. 5 Feet Wide; 4 Feet Deep.

INTERIOR OF BROODER HOUSE. FOR DAY-OLD CHICKS UP TO EIGHT WEEKS OLD.

180-Foot House, Capacity 400 Hens. Interior View Showing Glass Fronts on Left, Partitions and Dry-Mash Boxes on Right. A Typical Pullet in Foreground (White Leghorn) and One Feeding in the Dry-Feed Box. Scratching Litter also Shown.

180-Foot House (Back), Capacity 400 Hens, Showing North Side of Yard in Summer Time.

180-Foot House for 400 Birds. Southern Exposure.